A CONCISE SUMMARY OF NATE

The Signal and the Noise

...in 30 minutes

A 30 MINUTE EXPERT SUMMARY

GARAMOND
—PRESS—

CONTENTS

INTRODUCTION

Overview

According to Nate Silver, author of *The Signal and the Noise: Why So Many Predictions Fail – but Some Don't*, the information age has given people access to so much data that they're having real trouble making accurate predictions.

In the first half of the book, Silver describes how predictions can fail. For example, when people's political or personal biases interfere with how they receive information, their predictions often amount to little more than overconfident guesswork, since they're basing them not on the truth (the *signal*) but on what distracts them from the truth (the *noise*). People's failed predictions can even have catastrophic impacts on their own and others' economic and social well-being.

In the second half of the book, Silver shows how people can correct the inherent biases in their thinking and improve their prediction-making abilities in all areas. People can do this, he says, by applying Bayes's theorem to the task of making predictions. Bayes's theorem, Silver explains, is a stunningly simple and yet surprisingly rich mathematical formula that can guide everyone toward the kind of probabilistic thinking most likely to produce safer, more reliable predictions.

About the Author

Nate Silver (born January 13, 1978) is a statistician as well as an electoral and political analyst. His predictive abilities came to public attention through his stat-crunching activities in two highly visible American institutions—professional baseball and the arena of presidential elections.

When Silver was in his early twenties, he developed the PECOTA (Player Empirical Comparison and Optimization Test Algorithm) system for forecasting the performance and careers of Major League Baseball players. (The acronym is a private joke, since it spells out the last name of Bill Pecota, a pesky infielder with the 1980s-era Kansas City Royals who usually meant trouble for Silver's hometown favorites, the Detroit Tigers.) The PECOTA system proved so effective that it was purchased by the Baseball Prospectus organization.

In 2007, Silver began making predictions about the 2008 US presidential election, starting with the Democratic primaries and caucuses. At first he did this work under the pseudonym "Poblano" and published it on the Daily Kos political blog, but in March 2008 he founded his own site, FiveThirtyEight.com (named for the total number of electoral votes available in the fifty states). At FiveThirtyEight, Silver correctly predicted the winning presidential candidate in all but one of the fifty states in 2008, and he also accurately predicted the winners of all thirty-five US Senate races that year. In fact, with his nearly perfect predictions, Silver outperformed a range of better-known pundits and polls all through the 2008 electoral season. In the 2012 presidential election, Silver's notoriety skyrocketed when he correctly predicted the winning presidential candidate in all fifty states.

In 2009, *Time* named Nate Silver among the magazine's "100 Most Influential People" for that year. In 2010, Silver's blog was licensed to the

New York Times as FiveThirtyEight: Nate Silver's Political Calculus. In 2012, the International Academy of Digital Arts and Sciences honored FiveThirtyEight with a Webby Award as "Best Political Blog."

How the Book Came About

The Signal and the Noise, published in 2012, is Nate Silver's first book. Its publication is clearly an outcome of Silver's growing reputation as a statistician and political analyst. The book earned instant praise, as did Silver himself. The *Boston Globe* called Silver a "stats whiz" and deemed him responsible for remaking the art of political punditry. *Esquire* lauded *The Signal and the Noise* for bringing a "humanistic approach to statistics." The *Los Angeles Times* called the book "insightful," and the reviewer for the *Wall Street Journal* praised the scholarship of the author, describing Silver's arguments and examples as "painstakingly researched." Some critics have found the quality of the writing uneven, but the quality of the book's content has not been disputed. *The Signal and the Noise* was an instant success, and it quickly took its place among the top ten *New York Times* best sellers.

A CATASTROPHIC FAILURE OF PREDICTION

Overview

People's ability to generate data is growing all the time, Nate Silver says, but their prediction-making powers are not keeping up with the growth in data. Silver explains that confidence and precision, far from being signs of accuracy, can be indicators of *inaccuracy* when it comes to making predictions involving complex systems. Uncertainty, he says, is a part of life, and it's a big part of forecasting. But most forecasters fail to account for uncertainty, and that's why their predictions so often fail.

"Precise forecasts masquerade as accurate ones, and some of us get fooled and double-down our bets. It's exactly when we think we have overcome the flaws in our judgment that something as powerful as the American economy can be brought to a screeching halt."

– Nate Silver, *The Signal and the Noise*

Chapter Summary

For Nate Silver, catastrophically bad predictions by Wall Street ratings agencies and governmental institutions were at the heart of the US financial crisis of 2007–2008. Silver says that these predictions were bad because they all ignored key aspects of the contexts in which they were made. As one effect of this tendency, a panel of economists who participated in a December 2007 *Wall Street Journal* forecasting exercise made the confident prediction that there was only a 38 percent chance that 2008 would see a recession in the United States.

Silver also points to the decades-long US housing bubble, which finally collapsed in 2008, and to the variety of complex financial instruments that were created from collections of mortgage-backed securities. When the ratings agency Standard and Poor's (S&P) said that one category of complex mortgage-backed securities had only 1 chance in 850 of going into default, the agency miscalculated by a factor of 200. Silver claims that S&P, hoping to avoid responsibility for its poor statistical models, placed the blame on the housing bubble while also claiming not to have known anything about it—even though information about the housing bubble's impending collapse had appeared in news reports 3,447 times between 2001 and 2005, and even though additional documents show that S&P and other ratings agencies did indeed anticipate the housing bubble while ignoring its possible risks. In fact, their predictions about the housing bubble were based on what Silver calls irrelevant comparisons to the 1950s postwar US housing boom, whereas more relevant data was ignored. For example, some of the data that did not factor into the ratings agencies' accounts included information about the rates of default in Japan after the bursting of that country's real estate bubble in the 1990s.

In discussing the US housing bubble, Silver distinguishes *risk* (something on which a definite price can be put) from *uncertainty* (risk that is hard to measure). According to Silver, the ratings agencies performed a kind of alchemy that was designed to make uncertainty look and feel like risk. In other words, they claimed to know exactly how much risk was contained in what actually turned out to be unprecedented types of financial products, which were vulnerable to enormous amounts of systemic uncertainty.

In early 2009, several months after the September 2008 financial collapse and the resulting freeze-up of the credit markets, the new Obama administration made what Silver considers an inexcusable mistake. The administration boldly forecast the nature and characteristics of the post-stimulus economy without first preparing the public for the possibility of uncertainty, even though uncertainty is inherent in all macroeconomic forecasting. In this case, the administration ignored the possibility of uncertainty for the sake of appearing to make a confident, precise prediction—one that was ultimately wrong.

Silver uses the example of target practice to illustrate the difference between confidence and precision, on the one hand, and accuracy, on the other. You can know from long experience that your bullets always hit the target in the same place, and so you can be *confident* that your shooting is *precise*. But if the point is to hit the bull's-eye, and you never do, then you can't claim to be a good shot—that is, you can't say that your shooting is *accurate*.

When people make a prediction, Silver says, they often have their political, personal, and/or professional image on the line. But when they acknowledge uncertainty, their model may seem less rigorous, and their PowerPoint presentation may look less impressive. Nevertheless, Silver cautions, people should not allow an apparent lack of precision to discourage them from making better, more accurate predictions. Making better predictions requires people to acknowledge that they are biased, and that their thinking is always tainted by a subjective viewpoint.

Human beings, as Silver says, instinctively like confidence and precision, just as they instinctively *dislike* uncertainty, even though uncertainty is a part of life, and a big part of forecasting. But, Silver says, most forecasters continue to avoid expressions of uncertainty, just as key financial forecasters did several years ago when their confident predictions failed to warn Americans of the catastrophic financial crisis that loomed just ahead.

Chapter 1: Key Points

- When a prediction seems precise, and when it is put forward with boldness and confidence, it should be seen as inaccurate because it has failed to account for uncertainty.

- The 2007–2008 US financial crisis is one example of what can happen when statistical predictions fail catastrophically.

- Better statistical formulas—formulas accounting for uncertainty—might have made it possible to avoid the financial crisis of 2007–2008 and the recession that followed.

ARE YOU SMARTER THAN A TELEVISION PUNDIT?

Overview

Nate Silver, discussing the research of Philip Tetlock, notes that many experts who are specialists are so bad at making predictions that they might as well just toss a coin—and, Silver adds, Tetlock's research shows that the more information such experts acquire, the less accurate their predictions become. Silver attributes these specialists' failures to a cognitive style that lets their personal biases distort the facts. By contrast, Silver says, experts who are generalists fare much better as forecasters, because their cognitive style allows them to separate the world as it is from the world as they think it should be. He describes the generalists' cognitive style as characterized by thinking in terms of probabilities, adapting hypotheses to new information, and viewing a question from multiple perspectives, and he emphasizes the importance of recognizing and correcting for personal bias.

> *"Political experts had difficulty anticipating the USSR's collapse . . . because a prediction that not only forecast the regime's demise but also understood the reasons for it required different strands of argument to be woven together. There was nothing inherently contradictory about these ideas, but they tended to emanate from people on different sides of the political spectrum."*
>
> – Nate Silver, *The Signal and the Noise*

Chapter Summary

In this chapter, Nate Silver describes how a maxim attributed to the ancient Greek poet Archilochus—"The fox knows many little things, but the hedgehog knows one big thing"—became a central metaphor in research conducted by Philip Tetlock, a psychologist and political scientist. Tetlock studied how the forecasting ability of two sets of political experts was influenced by their thinking skills. He classified his experts along a spectrum that had specialists ("hedgehogs") at one end and generalists ("foxes") at the other, and he published his findings in a 2005 book, *Expert Political Judgment*.

Hedgehogs, according to Silver, are type A personalities whose faith in "big ideas" is as inflexible as it is unjustified. Hedgehogs have spent their careers bringing a narrow focus to bear on a few big ideas in their fields of expertise. Nate Silver describes hedgehogs' thinking style as stubborn, order-seeking, confident, and ideological. When hedgehogs encounter new information, he says, they use it to adjust their existing models, and when hedgehogs' predictions turn out to be wrong, they simply blame bad luck or outside forces. Silver notes that hedgehogs, such as television pundits, can be highly entertaining. As forecasters, however, they can't claim that their successful predictions are due to anything other than random chance.

In the realm of forecasting, Silver says, it's the generalists, or foxes, who excel. Foxes draw data from multiple fields. They approach problem solving from numerous angles, and they adapt their opinions as necessary. Silver describes the foxes' thinking style as tentative and empirical. Foxes qualify their positions, he says, and they take responsibility for their mistakes. Foxes also tolerate complexity as well as uncertainty, and they accept that some problems simply may not have solutions.

Silver identifies three primary characteristics of fox-like thinking—the ability to be *probabilistic*, the capacity for being *adaptable*, and openness to using *multiple perspectives*—and he says that his own political forecasting model, as illustrated by his FiveThirtyEight blog, incorporates these characteristics. He acknowledges that no one can be purely objective, but he affirms that accurate forecasting depends on the ability to notice and question one's personal biases.

Chapter 2: Key Points

- Experts can be divided into two camps—specialists ("hedgehogs") and generalists ("foxes"). Hedgehogs are often very bad at forecasting because their biases color their perceptions of the data. Foxes take uncertainty into account along with their personal biases.

- Effective forecasters don't limit themselves to quantitative data. They also use qualitative information, and they maintain awareness of their personal biases.

3

ALL I CARE ABOUT IS W'S AND L'S

Overview

Statistical forecasting models have been used in Major League Baseball for more than a decade, but they haven't replaced baseball scouts. Instead, Nate Silver says, the models have contributed quantitative data to the qualitative information that baseball scouts have always collected. Innovative scouts have discovered that access to more information helps them make better predictions. In the development of Major League Baseball teams, forecasting models now work in tandem with scouting, and the best results come from the most innovative thinkers.

"If Prospect A is hitting .300 with twenty home runs and works at a soup kitchen during his off days, and Prospect B is hitting .300 with twenty home runs but hits up nightclubs and snorts coke during his free time, there is probably no way to quantify this distinction. But you'd sure as hell want to take it into account."

– Nate Silver, *The Signal and the Noise*

Chapter Summary

More than a decade ago, forecasters using statistical models to make predictions shook up the world of Major League Baseball—starting with the baseball scouts, who worried that a bunch of stats geeks might be poised to throw them out of their jobs. Nate Silver's own innovative model, known as PECOTA, emerged during this period, an era depicted in the film *Moneyball*, based on Michael Lewis's book of the same title, which tells how Billy Beane, general manager of the low-payroll Oakland Athletics, used statistics to build a winning team and a victorious 2002 season.

Any forecasting system in baseball has to account for a player's statistics, determine how much of the player's performance has to do with blind luck, and adjust for how the player's performance alters as the player ages. PECOTA was designed to predict any one player's performance by comparing his statistics with those of earlier, similar players, using data culled from ten thousand player seasons. Silver's PECOTA system did miss a few calls, but it also scored some impressive hits.

Nevertheless, Silver reports, in the contest between PECOTA and the baseball scouts, it was the scouts who actually made the more accurate predictions. What made the scouts' predictions superior, he explains, was their access to more information than just stats—the availability of qualitative information in addition to quantitative data. According to Silver, when good forecasters get their hands on more information, their predictions improve (unlike the predictions of hedgehogs, which get worse and worse as more and more information becomes available). Therefore, Silver says, the computer-based forecasting models gave the scouts statistical data that they could use in tandem with such hard-to-quantify information as a player's habits or the range of his mental capacity.

Silver speculates that the competition presented by statistical models actually improved the baseball scouts' performance by challenging them to look past such biases as the one that overvalues a player's physical characteristics. (Because of that bias, scouts were dismissing Boston Red Sox star second baseman Dustin Pedroia while PECOTA was signaling his huge potential. Incidentally, the title of this chapter—"All I care about is W's and L's," or wins and losses—is a verbatim quote from Pedroia.) Silver says that good predictions, in baseball and other areas, require constant innovation as well as the ability to think both big and small in the quest for information that just may turn out to have value.

Chapter 3: Key Points

- Forecasting success in a competitive business like baseball depends on constant innovation. Good innovators think big, Nate Silver says—but they also think very small, since new ideas can be found in a problem's tiniest details, where no one else has even considered looking.

- If access to more information isn't improving a forecaster's predictions, then that forecaster has personal biases and/or a problematic cognitive style to deal with.

- Computer-based forecasting models seem to have improved the performance of Major League Baseball scouts, probably by challenging their biases.

FOR YEARS YOU'VE BEEN TELLING US THAT RAIN IS GREEN

Overview

In this chapter—the title cites viewers' alarmed reaction after the Weather Channel abruptly changed its graphics to represent rain with blue instead of green shading—Nate Silver explains why weather forecasts have been growing more accurate, even though the weather is intrinsically difficult to predict.

> *"Accuracy is the best policy for a forecaster. It is forecasting's original sin to put politics, personal glory, or economic benefit before the truth of the forecast."*
> – Nate Silver, *The Signal and the Noise*

Chapter Summary

Nate Silver says that the accuracy of weather forecasts has improved 350 percent in the last twenty-five years, in great part because of computers. But the weather is hard to forecast because it's a complex system—that is, a system that is both dynamic and nonlinear. With an ever-changing system like the weather, Silver says, forecasts are difficult not only because the system is characterized by chaos, or inherent unpredictability, but also because even the tiniest input error can produce exponentially large outcome effects.

Silver explains that accurate weather forecasts also require input from the human eye, which can quickly observe what a computer cannot. In fact, he says, human judgment can improve forecasts of rain by 25 percent and temperature forecasts by 10 percent.

But human judgment, for various reasons, also plays a role in making forecasts less accurate. In 2005, for example, the National Hurricane Center's forecasts concerning Hurricane Katrina were accurate and timely enough for New Orleans to have been evacuated before the storm surge breached the city's levees. But the mayor failed to make evacuation mandatory, Silver says, and some 1,600 people lost their lives.

For Nate Silver, perhaps the most important test of a forecast is *calibration*—over the long run, if it actually did rain about 40 percent of the time for every time you predicted a 40 percent chance of rain, then your forecasts were well calibrated. But calibration depends on probabilistic thinking, something Silver says most people aren't very good at.

Chapter 4: Key Points

- The weather is a system that is both dynamic and nonlinear, and so the weather is intrinsically difficult to forecast. Even so, the accuracy of weather forecasts has improved greatly over the past twenty-five years.

- Public weather-forecasting agencies typically provide the most accurate forecasts, whereas commercial weather forecasters tend to sacrifice real accuracy for the "added value" of *perceived* accuracy.

(5)

DESPERATELY SEEKING SIGNAL

Overview

As Nate Silver explains in this chapter (the title is a play on *Desperately Seeking Susan*, a 1985 film costarring Madonna), particular earthquakes cannot be predicted, at least not in the sense of forecasting the dates and times when they will occur. All that seismologists can do is use statistics to establish parameters for how probable earthquakes are in a particular region, and for how frequent they are likely to be.

> *"If you have hundreds of people trying to make forecasts, and there are hundreds of earthquakes per year, inevitably someone is going to get one right."*
>
> – Nate Silver, *The Signal and the Noise*

Chapter Summary

Geological systems, like weather systems, are *inherently chaotic*, Nate Silver says—that is, they're characterized by unpredictability. But as little as people really know about weather systems, they know even less about what's going on in the unobservable underground geological systems that produce earthquakes. An earthquake may also belong to an *inherently complex* system, one marked by long periods of stasis that are periodically and catastrophically upset. If so, Silver says, the ability to predict earthquakes is even further beyond people's reach, and seismologists must rely on statistics alone, not to predict earthquakes but only to establish parameters for their probability and frequency.

Because earthquake-related data is "noisy" and limited, Silver explains, and because seismologists have a poor understanding of the relationships within earthquake systems, conditions are ripe for a statistical error called *overfitting*. This error takes the form of providing an *overly specific* solution to a *general* problem. As one example, Silver cites the tsunami that inundated Japan's Pacific coast—and the Fukushima Daiichi Nuclear Power Plant—after a March 2011 earthquake measuring 9.1 on the Richter scale. That quake was five times larger than the 8.6 earthquakes that the nuclear plant had been built to withstand, and the tsunami's 130-foot waves also proved too high for the structure to handle. At the time of the plant's construction, archaeological evidence had been available to suggest a history of tsunamis capable of producing such huge waves. Apparently, however, that evidence of earlier devastation was either forgotten or ignored. Overfitting can be an honest mistake, Silver says, but sometimes it's deliberate, as when planners, for the sake of producing successful-looking models, give more weight to the noise (distraction) of their own rationalizations than to the signal (truth) of actual data.

According to Silver, irresponsible predictions of earthquakes disrupt people's lives, spread misinformation, and impede the progress of seismological science. These are three reasons why the seismological community does not get involved in issuing earthquake predictions.

Chapter 5: Key Points

- The location and the date of a particular earthquake cannot be predicted, but seismologists can use statistical modeling to forecast probabilities for how frequently earthquakes will strike a particular area.

- Whether by accident or design, earthquake forecasting models are susceptible to an error known as *overfitting*, whereby an overly specific solution is provided for a general problem.

HOW TO DROWN IN THREE FEET OF WATER

Overview

The title of this chapter refers to a catastrophic 1997 flood in Grand Forks, North Dakota, where forecasters' fear of appearing uncertain led them to misstate, by a mere three feet, the capacity of the town's levees to hold back the rising Red River of the North. But the chapter itself is about economic forecasting, which Nate Silver says is difficult at best.

"An oft-told joke: a statistician drowned crossing a river that was only three feet deep on average."

— Nate Silver, *The Signal and the Noise*

Chapter Summary

As Nate Silver argues in this chapter, an economy is a complex system to begin with. Moreover, economic forecasters face the daunting challenge of trying to separate causes from effects when causes and effects are not only intertwined but also susceptible to feedback loops that depend on unpredictable, emotional human behavior, such as what is seen during panics and bubbles. Economic forecasting also suffers from forecasters' bias toward overconfidence as well as from their fear of looking bad if they acknowledge uncertainty, which leads them to let their reputations take precedence over accuracy.

But Silver says that economists should always include uncertainty in their predictions, and that they should never commit the error of seeing causal relationships where none exist. This is known as the error of confusing *correlation* with *causation*. To illustrate the difference, Silver uses the example of forest fires and ice cream sales. These two phenomena are statistically related, since they both spike during the summer, but it would be absurd to conclude that one causes the other. Nevertheless, he says, this kind of error is very common in economic forecasting, and so is the error of ignoring data that doesn't support the forecaster's existing model.

Silver believes that the entire system of economic forecasting needs to be overhauled, because forecasters working in the current system are often motivated less by making accurate predictions than by looking good to their colleagues. As a result, they lack incentives for overcoming their biases. He cites evidence that anonymous forecasters tend to make more accurate predictions than those whose names and reputations are known. On that basis, he seconds the economist Robin Hanson in suggesting that a market-based approach—such as the use of prediction markets, where bets are made on future outcomes—might improve the accuracy of predictions by attaching forecasts to financial stakes. That kind of shift, he says, might increase demand for accurate economic forecasts and reduce demand for forecasts that are overconfident and wrong.

Chapter 6: Key Points

- In any forecast, uncertainty is a better indicator of accuracy than confidence is.

- Economic forecasters work with challenging data in an ever-shifting, complex environment, and yet they routinely sacrifice accuracy, cover up uncertainty, and make confident-sounding predictions just so they can look good to their peers.

- When forecasters ignore data, it's a clear sign that they'd rather show off than make accurate predictions.

ROLE MODELS

Overview

As Nate Silver explains, accurate prediction of the outbreak and spread of an infectious disease is a matter of life or death. As a result, medical statisticians have no incentives for posturing, but they do have every reason to correct for their own distortion of the signals they're receiving. According to Silver, that's what makes these statisticians among the most thoughtful and honest forecasters in any field.

> *"Because of medicine's intimate connection with life and death, doctors tend to be appropriately cautious. In their field, stupid models kill people."*
> – Nate Silver, *The Signal and the Noise*

Chapter Summary

In 1976, public health forecasters erroneously predicted a deadly US epidemic of swine flu (H1N1) and pressed President Gerald Ford to rush a nationwide vaccination program into operation. As Nate Silver reports, the epidemic never materialized, but a significant statistical anomaly did occur—some five hundred people who received the vaccine came down with Guillain-Barré syndrome, a rare and very serious autoimmune disorder.

Prediction models can be flawed in numerous ways, Silver says. They can create both *self-fulfilling* and *self-canceling* predictions. They may extrapolate from data that can be useful only after an outbreak has run its course, or they may take people's reports of symptoms at face value when such reports merely reflect obsessive media coverage of an outbreak, to the point where it's not clear whether a disease is truly spreading or just more prominent in the news.

Simplicity in a model can be a good thing, Silver says—a model that's too complicated can pick up too much noise and misrepresent the underlying structure of a problem. But simplicity can be a flaw when it's not paired with adequate sophistication.

As an example of one solution to such flaws, Silver describes *agent-based modeling*, a technique now being developed to predict the spread of the staph infection commonly known as MRSA (*Methicillin-resistant Staphylococcus aureus*). This technique is so complex that it requires the support of supercomputers. Agent-based modeling takes account of very granular details, such as how many people in a neighborhood have been to prison (where staph infections thrive) and the culture-based frequency of hugging (a means of transmitting the infection). As sophisticated as this technique is, however, it still lacks data, Silver says—there's just not enough. This is partly because agent-based modeling anticipates trends that have not yet appeared, and partly because the rarity of certain diseases makes agent-based modeling hard to test.

Chapter 7: Key Points

- Nate Silver sees US medical statisticians as unusually thoughtful and honest. Nevertheless, over the past four decades, their use of flawed models has caused them to fail significantly in predicting the outbreak and spread of infectious diseases like swine flu (H1N1) and AIDS.

- *Agent-based modeling* is a developing and exceptionally ambitious approach to predicting outbreaks of infectious diseases. This technique uses multidisciplinary teams of experts. But even with all their focus and intelligence, these experts often find their efforts frustrated by lack of data.

- Even the most effective prediction model will have no impact on a problem whose solution depends on how well forecasters understand themselves and how accurately they account for their interpretations and distortions of the signals they receive.

8

LESS AND LESS AND LESS WRONG

Overview

At the beginning of this chapter, Nate Silver explains that the title was inspired by the Danish mathematician and poet Piet Hein, who characterizes the "road to wisdom" as learning to "err and err and err again, but less and less and less." Silver goes on to describe R. A. Fisher's frequentist approach to statistics, a twentieth-century method that replaced the theorem developed by Thomas Bayes, an eighteenth-century English minister and philosopher. The frequentist approach starts from the premise that humans are capable of unbiased reasoning and that knowledge exists in an objective sphere separate from the actual world. By contrast, Silver explains, an approach based on Bayes's theorem is more effective at separating the signal of data from the noise of distractions, such as the bias inherent in all human thinking. In Silver's opinion, statistical science took a step back when frequentist methods replaced Bayes's theorem, but he thinks that a welcome return to Bayesian methods may now be under way in statistical science.

22

> *"Finding patterns is easy in any kind of data-rich environment; that's what mediocre gamblers do. The key is in determining whether the patterns represent noise or signal."*
>
> – Nate Silver, *The Signal and the Noise*

Chapter Summary

According to Nate Silver, Thomas Bayes observed that probabilistic beliefs get formed when human beings are confronted with new information about the world. In that situation, people make mental *approximations* of reality, revising them whenever new evidence is gained, and with each approximation, people move closer to the truth. For Bayes, thinking itself was a probabilistic endeavor.

At its most basic, Bayes's theorem is simply an algebraic expression with three known variables and one that is unknown. To use it in testing a hypothesis about an observed phenomenon, three elements are necessary: (1) an estimate of how probable it is that the hypothesis is true, (2) an estimate of how probable it is that the hypothesis is false, and (3) a prior probability, or the probability that would have been assigned to the existence of the observed phenomenon before it was actually observed.

Silver says that as long as frequentist methods dominated twentieth-century statistical science, there were high rates of false positives as well as high failure rates for statistical predictions. As evidence for this claim, he cites a 2005 paper by John P. A. Ioannidis, "Why Most Published Research Findings Are False," a study that illuminates the flawed statistical and theoretical foundations that have made predictions worse in the *age of Big Data*. Silver even says that exponential growth in data production has done little more than magnify errors in thinking. Human thinking cannot attain objective purity, he says, because people *approximate* their way toward truth.

Chapter 8: Key Points

- Bayes's theorem assumes that bias is inherent in the human mind, and so it requires people to think probabilistically from the very start. In this way, it increases people's ability to discern the signal of data from the noise of bias and other distractions.

- The frequentist approach considers statistical errors to be errors of measurement. The Bayesian approach, which deals with epistemological uncertainty, views the limits of human knowledge as the source of statistical errors.

RAGE AGAINST THE MACHINES

Overview

In this chapter, Nate Silver uses the three-part structure of a game of chess, in addition to several entertaining accounts of man-versus-computer chess matches, to make an overarching point: that computers do only what they are programmed to do. In addition, he discusses the ways in which human beings, without the aid of supercomputers, are able not only to make Bayesian predictions but also to test and refine those predictions in real-life contexts.

"Whatever biases and blind spots the forecaster has are sure to be replicated in his computer program."
— Nate Silver, *The Signal and the Noise*

Chapter Summary

According to Nate Silver, a disjunction has occurred between human evolution and technological progress. Evolution, he says, takes place over thousands of years, but the processing power of computers doubles approximately every twenty-four months.

One result of this disjunction, Silver says, is that people are simply not capable of making *perfect* decisions in the face of vastly greater amounts of information than any human being can ever hope to process at any one time. The best people can expect of themselves, he says, is to make the *best* possible decisions, specifically by acknowledging their human limitations, in keeping with Bayes's theorem.

Computers, too, have limitations, Silver notes. For example, they can't think creatively or see the big picture. Computers cannot make accurate predictions within complex systems, and so they are unreliable forecasters of economic trends or events like earthquakes. When computers are most useful—as they are in fields like weather prediction and chess—it's because the context is one in which a system operates by relatively simple and well-understood laws, and because good forecasting depends on the ability to solve repetitive equations much more rapidly than a human being can.

Chapter 9: Key Points

- Within simple, well-understood systems, computers can outperform humans at certain tasks. For example, they excel at making fast computations, tirelessly and unemotionally. When they make errors, it's only because they've been programmed to do so.

- Both the individual human brain and the supercomputer can be useful in testing and correcting hypotheses in real life—which, according to the Bayesian theorem, is the only place where knowledge can actually exist.

THE POKER BUBBLE

Overview

Nate Silver frames this chapter in terms of the *poker boom* that began around 2003. The resulting *bubble economy* attracted legions of poker players who were so inexperienced that even marginally more competent players were reaping huge profits because the *water level* of the competition was so low. Poker, as Nate Silver describes it, is a highly mathematical game whose best players use Bayesian methods both to make predictions and to constantly hone their predictive skills.

> *"In politics, I'd expect that I'd have a small edge at best if there were a dozen clones of FiveThirtyEight. But often I'm effectively "competing" against political pundits, like those on The McLaughlin Group, who aren't really even trying to make accurate predictions. Poker was also this way in the mid-2000s. The steady influx of new and inexperienced players who thought they had learned how to play the game by watching TV kept the water level low."*
>
> – Nate Silver, *The Signal and the Noise*

Chapter Summary

For Nate Silver, the game of poker is a process that is Bayesian to the core. This is what makes it such a practical example of how mathematical skill can be used to make probabilistic judgments in the face of constant uncertainty.

According to Silver, it's math, not some mystical ability to read another player's "poker face," that allows a skilled player to make probabilistic forecasts of the cards in other players' hands. Good players, he says, think in terms of conditional probabilities. They mentally calculate and weigh numerous possibilities, and their calculations also include those possibilities that their opponents may be calculating. For Silver, nothing beats Bayesian methodology for knowing when to hold 'em and when to fold 'em.

In all fields and across disciplines, Silver says, sometimes the point is not how good your predictions are in an absolute sense but rather how good they are as compared to those of the competition. In other words, you can be making accurate predictions 95 percent of the time, but that won't matter much if your competitors are nailing theirs 99 percent of the time.

And so, Silver explains, sometimes the best thing is to focus less on the *results* of forecasting and more on the *process* of forecasting. In other words, he explains, if there's so much noise in a forecaster's sample of predictions that people can't tell whether those predictions are any good over the long run, they can ask themselves whether that forecaster is applying the attitudes and aptitudes known to be associated with successful forecasts over the long run. In this way, Silver explains, people are actually making a prediction of sorts about how good that forecaster's predictions are likely to be.

Chapter 10: Key Points

- Uncertainty is an inevitable part of probabilistic thinking.

- Poker is a mathematical game in which good players, faced with ever-present uncertainty, demonstrate a high degree of skill in probabilistic forecasting.

- When a data set is noisy, as in a game of poker, it can be advantageous for players to focus more on the *process* rather than on the *outcomes* of their probabilistic forecasts. For experienced players—as users of Bayesian methods, and therefore as self-improving forecasters—that process includes having learned to be humble, to avoid overconfidence, and to avoid mistaking blind luck for personal forecasting prowess.

IF YOU CAN'T BEAT 'EM . . .

Overview

In the stock market, Nate Silver says, Bayesian process can help people take advantage of group forecasting, which is useful in market predictions. But even though an aggregated (group) prediction, if made correctly, is better than an individual prediction, it does not necessarily follow that *every* aggregated forecast is *better* than *any* individual's *best* forecast. An investor's best hope, Silver says, is to avoid herd behavior and learn to detect stock market bubbles in real time.

> *"We'll never detect a bubble if we start from the presumption that markets are infallible and the price is always right. Markets cover up some of our warts and balance out some of our flaws. And they certainly aren't easy to outpredict. But sometimes the price is wrong."*
>
> – Nate Silver, *The Signal and the Noise*

Chapter Summary

Stock market predictions are never perfect, because they reflect fallible human judgment. But if people use Bayesian thinking in the stock market, they can adjust their probabilistic beliefs as they encounter new information.

Nevertheless, Silver warns, it is important not to become overconfident in personal forecasting abilities. Group forecasts are better than individual forecasts, he explains, when individual forecasts are made independently, before aggregation. But this doesn't mean that a group forecast is necessarily better than the best individual forecast.

Prices in the stock market can be predictable over the short run but not over the long run. Silver explains this phenomenon by pointing to the behavior of stock traders, which has its own impact on the market. For example, if a trader buys and the market goes up, then the trader thrives personally and professionally. If the trader sells and the market drops, then the trader is a genius in that scenario, too. If a trader follows the herd and buys, and the market crashes, then the trader's position becomes tenuous. But if the trader, refusing to follow the herd, decides to sell and the market rises, then that trader's career is ruined, and no one is likely to hire that trader again.

Silver cites evidence showing that herd behavior is common among traders on behalf of institutional investors. Traders are judged by their short-term performance, and this is just the type of incentive that leads to market bubbles. Silver believes that bubbles will always occur in the economic system. But even though bubbles can't be prevented, Silver argues, people can hope to become better at detecting them while they're inflated.

Chapter 11: Key Points

- Aggregate market predictions are generally better than individual predictions. Nevertheless, individual and aggregate predictions are equally susceptible to fallible human judgment, and so there is no aggregate prediction that is *necessarily* better than the *best* individual prediction.

- Traders are judged on the basis of short-term performance. That fact, combined with the fact of fallible human judgment, creates an incentive for traders to follow the crowd in order to protect their income. As a consequence of herd behavior like this, market bubbles are not just possible but probable.

- It takes discipline for an individual investor to avoid herd behavior, which means not buying during a bubble, when prices are soaring, and not selling during a panic, when prices drop. The best policy is to stick with investments over long periods when the *trend* is upward, regardless of momentary ups and downs in the market.

A CLIMATE OF HEALTHY SKEPTICISM

Overview

The debate over climate change, as viewed through a political lens, looks nothing like the same debate as viewed through the lens of the scientific community. But these two lenses have been superimposed, and the result, Nate Silver says, has been an acceleration of noise over signal.

> *"In the scientific argument over global warming, the truth seems to be mostly on one side: the greenhouse effect almost certainly exists and will be exacerbated by manmade CO_2 emissions. This is very likely to make the planet warmer. The impacts . . . are uncertain, but are weighted toward unfavorable outcomes."*
>
> – Nate Silver, *The Signal and the Noise*

Chapter Summary

According to Nate Silver, the current debate over climate change is an example of what can happen when science becomes crossed with politics—the result, he says, is more noise than signal. Silver claims that most climate scientists agree on the existence and current activity of the greenhouse effect, as intensified by human-generated emissions of carbon dioxide (CO_2). What they disagree on, Silver says, is the effectiveness of the computer models that are forecasting climate change as one result of these two phenomena. Although increases in temperature and rises in sea levels have been fairly well modeled, there is a great deal of uncertainty about changes that may still come as CO_2 levels exacerbate the greenhouse effect.

In the scientific realm, that kind of uncertainty is all in a day's work. But in the political realm, Silver says, uncertainty is equated with weakness. As a result, Silver claims, although some overconfident forecasts about climate change can be ascribed to modeling errors, others have been the results of deliberate attempts by politically engaged scientists to influence governmental actions on a global scale.

But, Silver reminds everyone, overconfident predictions about complex systems are always indicators of inaccuracy. Moreover, scientists are inclined to learn from failed forecasting models, in keeping with Bayesian principles. But political players who've seen poor forecasting models exposed as such have used those failures to mislead the public into discounting the greenhouse effect. As a result, Silver says, fewer Americans over the past several years have believed in the possibility of global warming.

The real problem, according to Silver, lies in the polarized political system, which does not reflect the intelligence of the American people. In politics, Silver explains, truth takes a backseat, and there is little incentive to separate signal from noise. Silver advises scientists to avoid the political realm, and to keep climate-change theory distinctly within the realm of science, where it is likely to be refined by other scientists who are using Bayesian methods.

Chapter 12: Key Points

- Most climate scientists now agree that the greenhouse effect is real, and that it is probably exacerbated by human-produced CO_2 emissions.

- Climate scientists disagree about some overconfident computerized models that have forecast climate change as a result of the greenhouse effect and CO_2 emissions. But certain people, including some scientists, have distorted and exploited that disagreement for political ends, spreading confusion and unwarranted skepticism about climate change, and setting climate science back in the process.

- The accuracy of scientific claims can probably be tested more effectively by scientists, whose goal is scientific truth, than by political actors, for whom truth is not the ultimate goal.

WHAT YOU DON'T KNOW CAN HURT YOU

Overview

For Nate Silver, the 2001 terrorist attacks on the World Trade Center and the Pentagon were unimaginable before they occurred. And yet, he notes, mathematical analysis indicates that the 9/11 attacks were entirely predictable as elements of a particular and known pattern. An effective defense against further terrorism, Silver says, requires people to get busy imagining even bigger attacks.

> *"When it comes to terrorism, we need to think big. . . . Signals that point toward . . . large attacks should therefore receive a much higher strategic priority."*
>
> – Nate Silver, *The Signal and the Noise*

Chapter Summary

In security intelligence, Nate Silver says, it is extremely difficult to detect signal from noise. Only after an attack has occurred does the signal leap out from the noise. Only then does it seem possible that the signal could ever have been buried.

The problem is the human bias for equating the strange with the implausible. But people can overcome this bias by intentionally imagining implausible events. To prevent future attacks, Silver explains, advance detection will require effortful imagination—thinking about the unthinkable, such as a nuclear or biological attack on the United States, with casualties that are exponentially greater than those of the 9/11 attacks.

Mathematics, as applied to terrorism, reveals indications of frequency and intensity, showing that the attacks of 9/11, instead of being outliers, belonged to a pattern, one in which a *power-law distribution* was at work. In this kind of pattern—the same pattern that characterizes earthquakes—a very small number of incidents (earthquakes or terrorist attacks) accounts for a greatly disproportionate amount of impact. In terms of this analysis, the scale of the 9/11 attacks should not have been beyond imagining, and people should be able to imagine even more massive attacks in the future.

In situations where decision makers face high uncertainty, Silver advises intelligence analysts to use the Bayesian approach—to think probabilistically about numerous hypotheses, routinely update them, and imagine possibilities that can reduce what Donald Rumsfeld famously called "the unknown unknowns" (in other words, what people have failed to imagine). For Silver, the greatest failure with respect to making a difficult prediction is to make no prediction at all.

Chapter 13: Key Points

- Mathematical analysis suggests that the terrorist attacks of September 11, 2001, were not an outlier.

- Terrorist attacks are like earthquakes in that they follow the pattern of a *power-law distribution*. In this pattern, a very small number of incidents account for a disproportionately great impact in the domain under investigation.

- Threats to security have escaped detection because the human brain automatically rejects the unfamiliar as unlikely. But terrorism, unlike earthquakes, can be prevented. Bayesian thinking, along with vigorous efforts to imagine the implausible, can aid the work of intelligence analysts and help to prevent further terrorist attacks on the United States.

CONCLUSION

Human beings are biased toward thinking that they are good at predicting things, Nate Silver claims. And yet, he says—perhaps especially in the age of Big Data, as Silver characterizes this era—people actually have a great deal of trouble making accurate predictions.

According to Silver, people can begin to improve their predictions by taking a Bayesian approach to forecasting. This approach, based on Bayes's theorem, requires people to recognize that, because of the way their brains are wired, they perceive truth by making approximations from simplified models of reality. The Bayesian approach also requires people to accept the expression of uncertainty as a necessary part of thinking.

People who are just beginning to make predictions may do poorly at first. But as they slow their thinking down and consider not only the data but also the personal biases that they bring to their analyses, they are likely to experience improvement in the skills that make up probabilistic thinking.

In fact, Silver reminds everyone, Bayes's theorem requires people to be honest right from the start about their beliefs in an event's likelihood. Flexibility is the key to Bayes's theorem—as people acquire new information, they update their forecasts and learn from their mistakes (something that Philip Tetlock's ideological hedgehogs are characteristically too proud to do). Finally, Silver says, if people want to distinguish the signal of truth from the noise of their biases and other distractions, they need not only knowledge of science but also ever-increasing knowledge of themselves.

CPSIA information can be obtained at www.ICGtesting.com
Printed in the USA
LVOW080435250113

317127LV00002BC/340/P